Cook

by Ann-Marie Kishel

first step nonfiction

Lerner Publications Company · Minneapolis

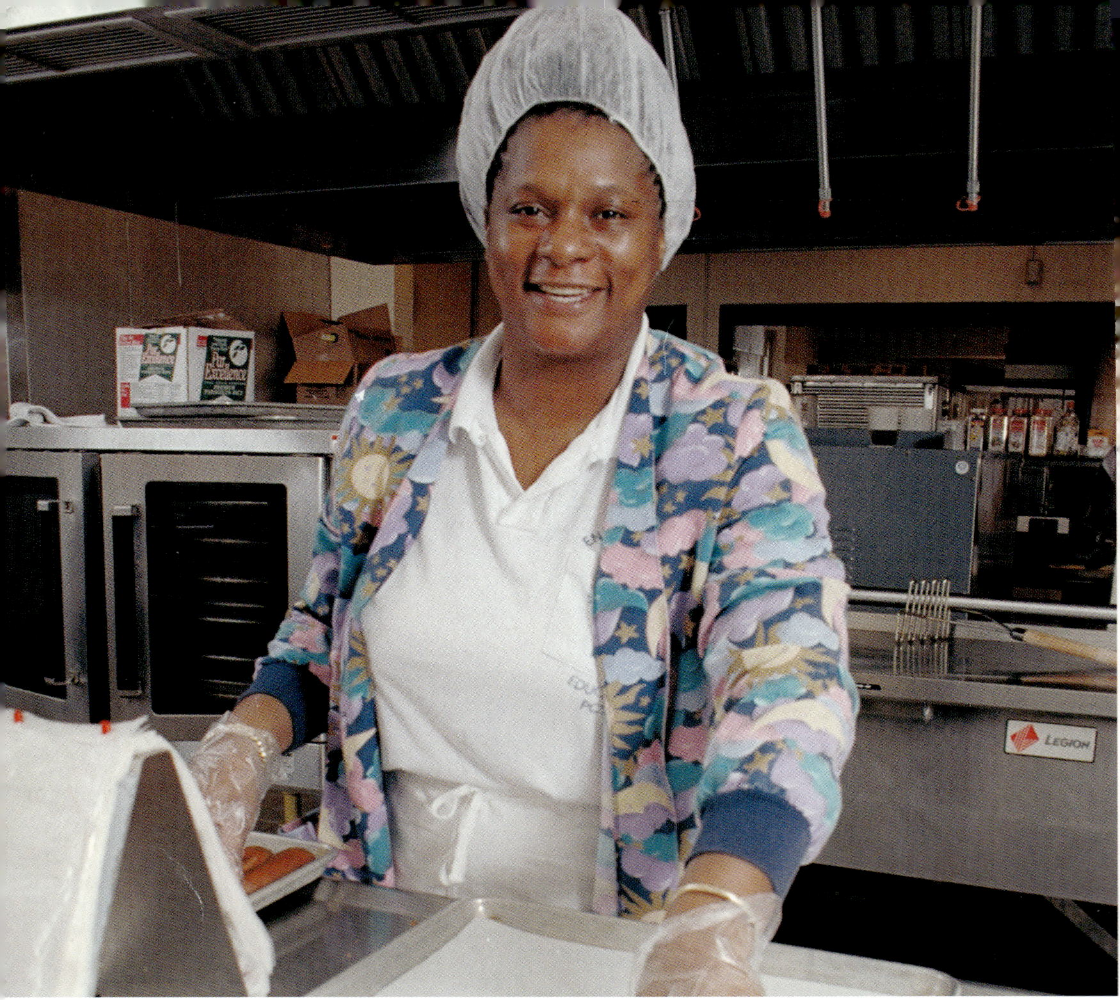

What does a cook do?

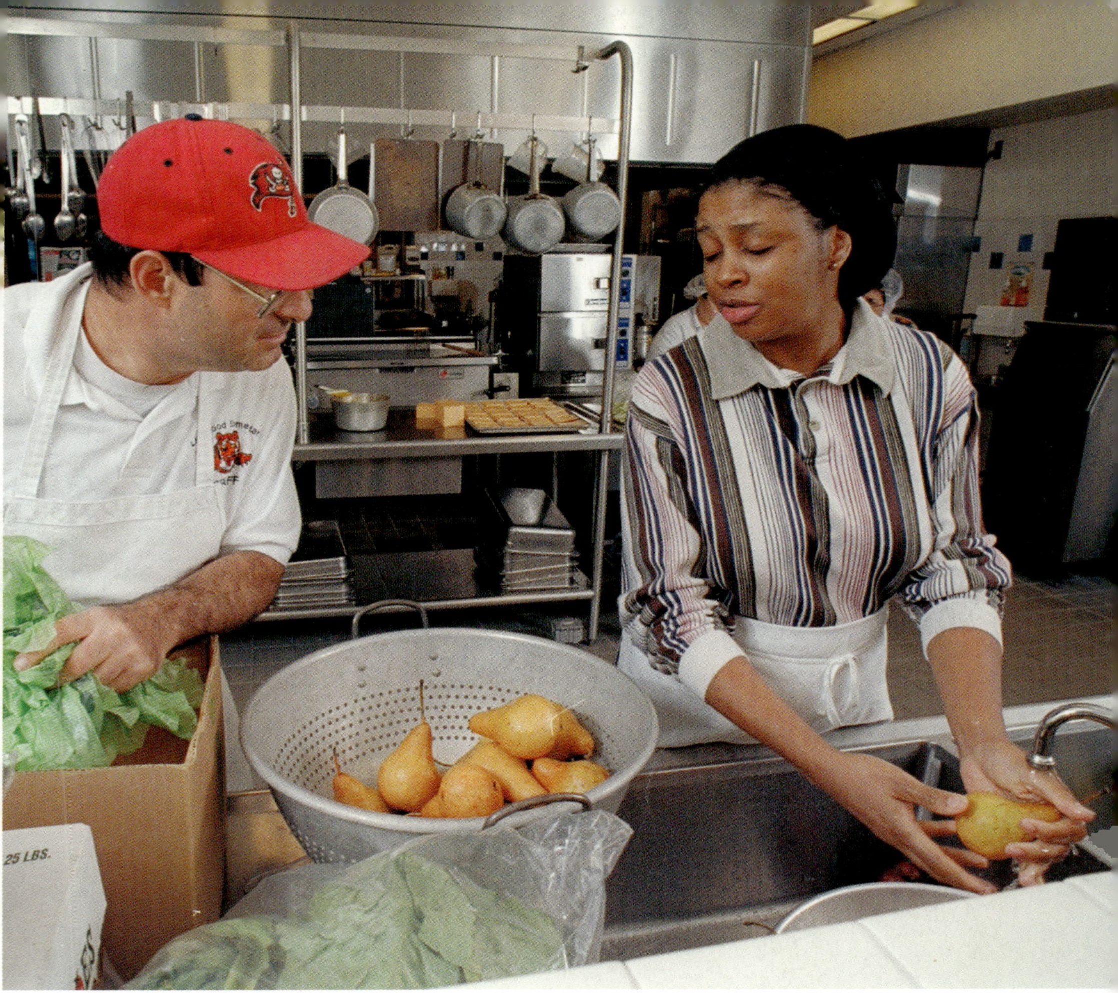

She washes.

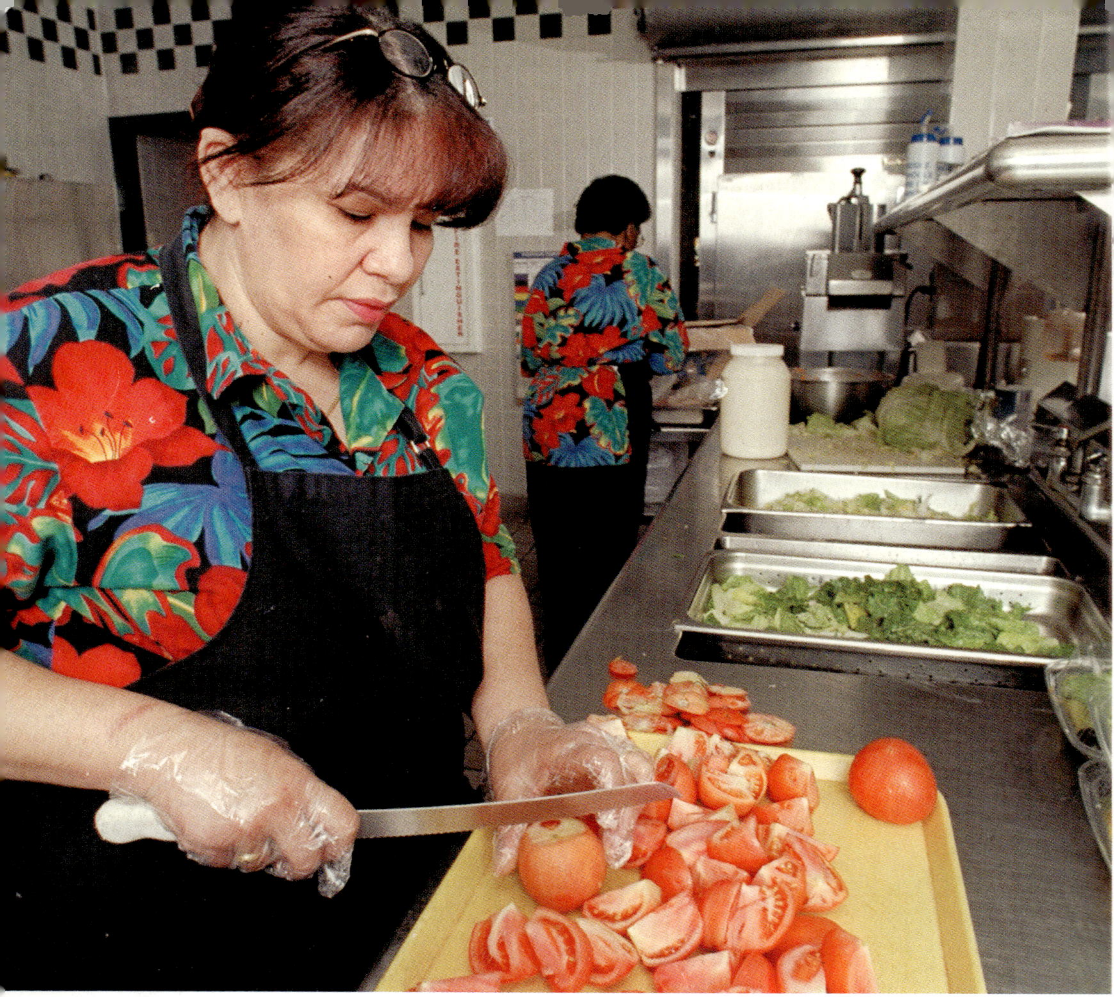

She cuts.

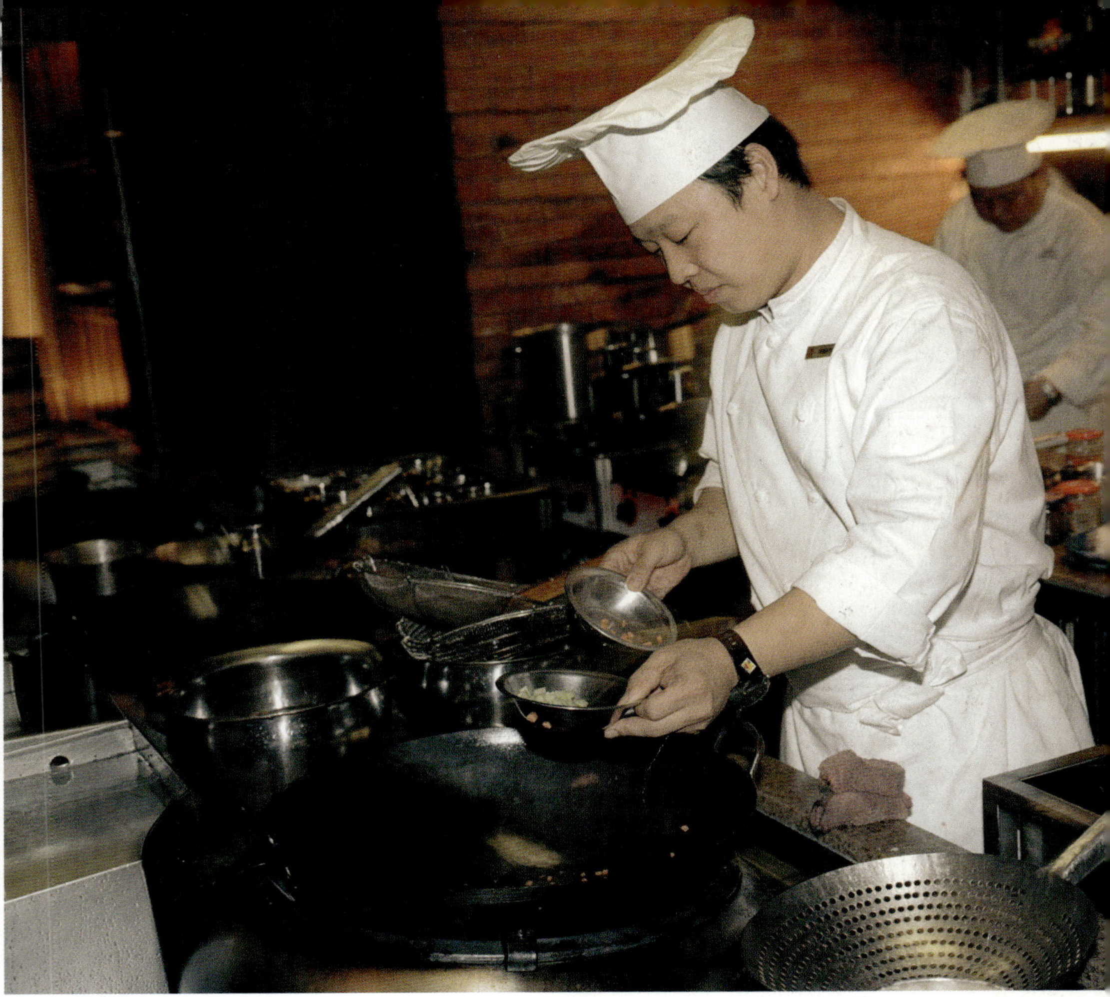

He mixes.

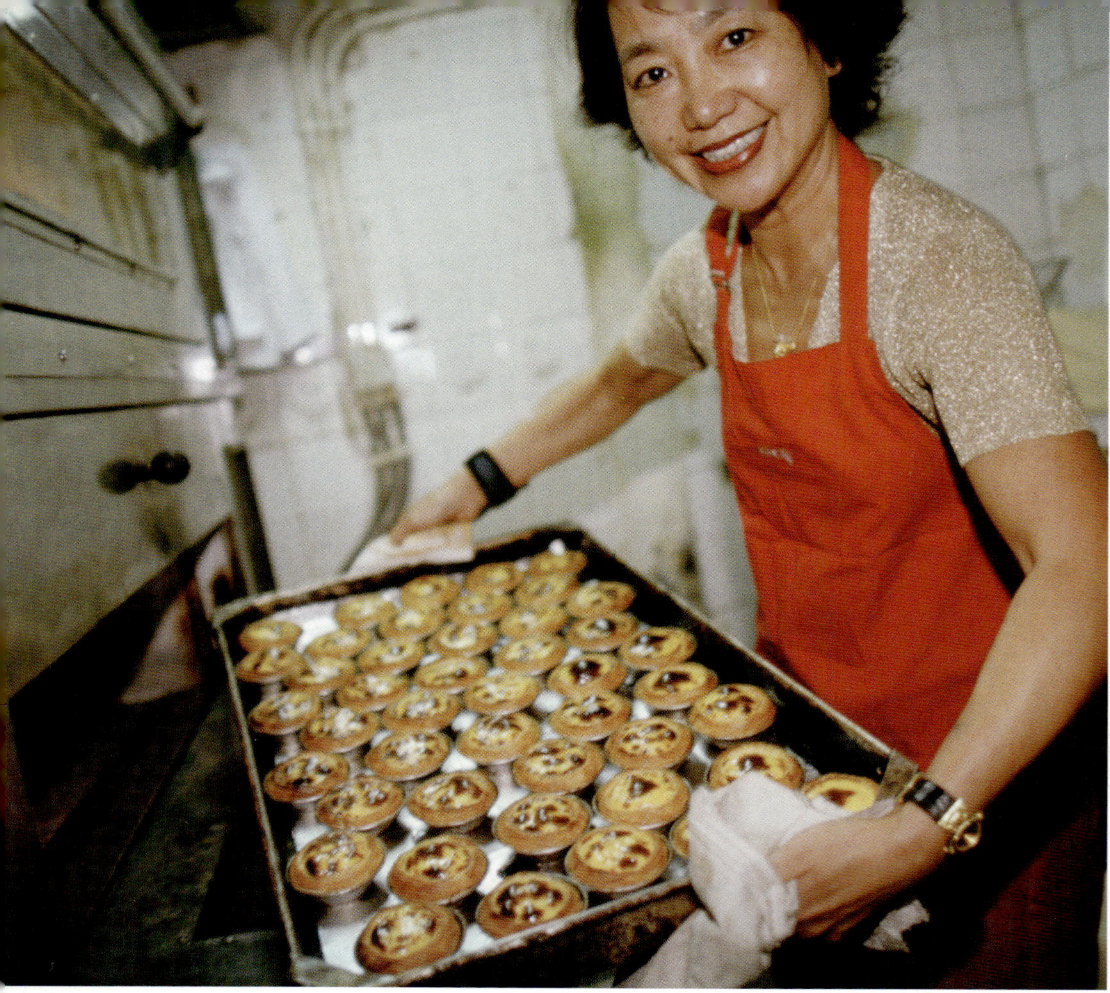

She bakes.

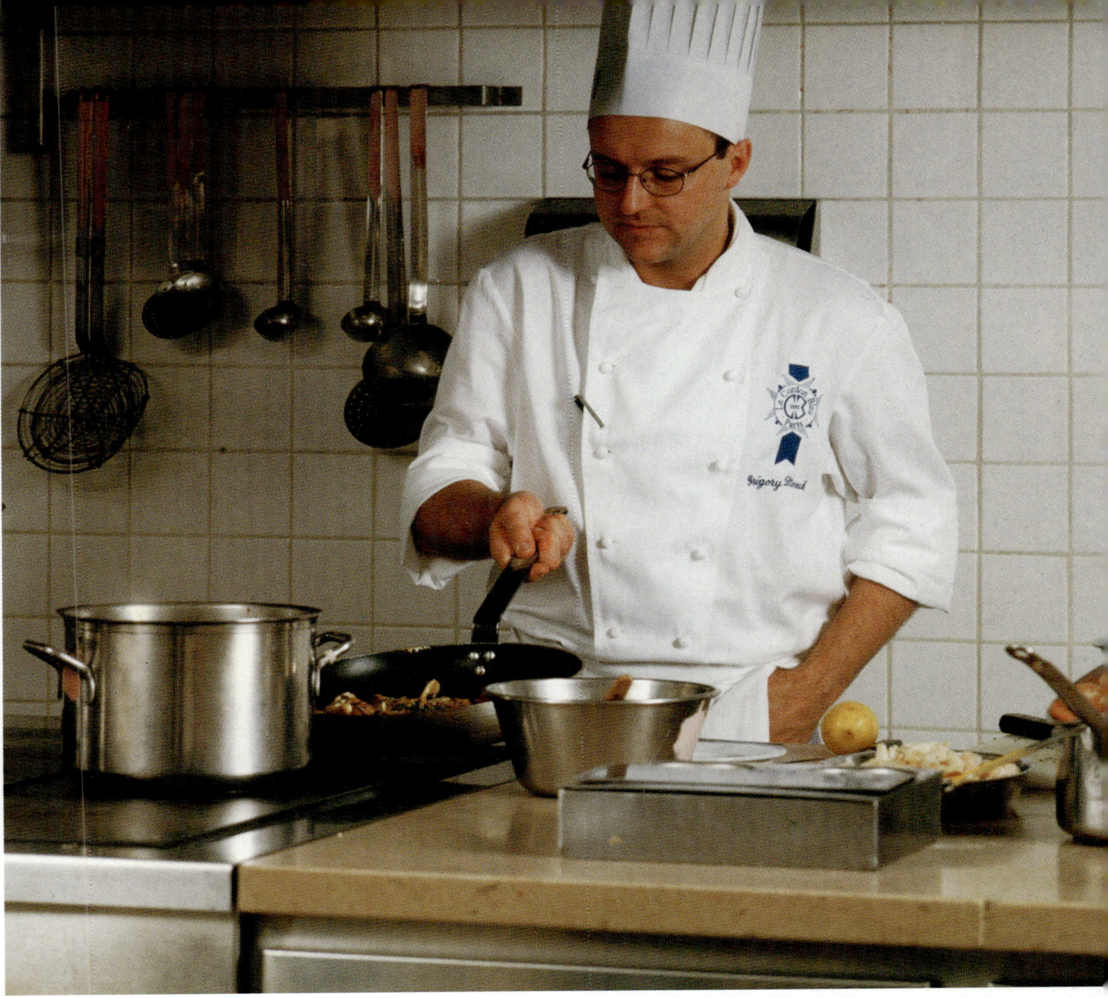

He fries.

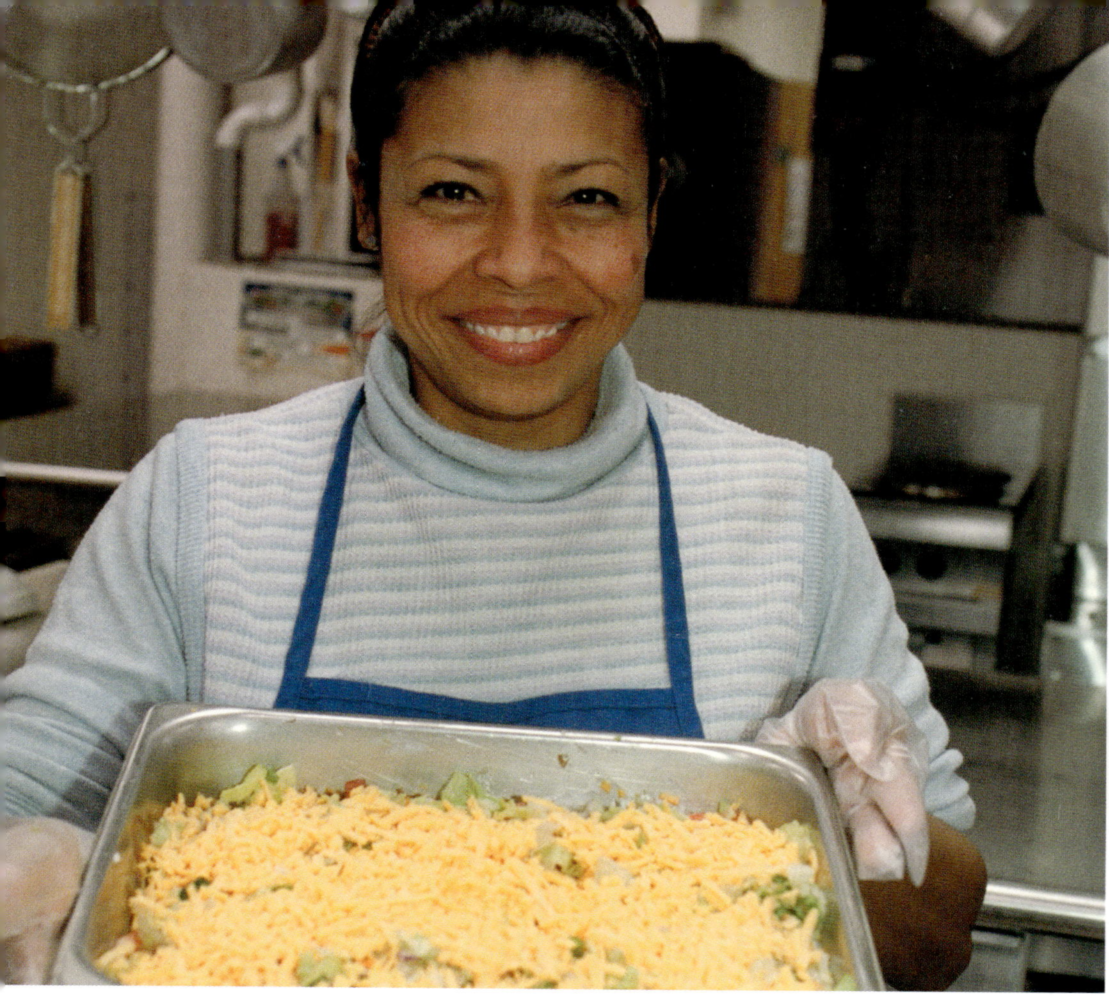

Do you know a cook?